Spotlight on Georgia
Performance Standards

HSP Georgia
Science

Interactive Text

Harcourt
SCHOOL PUBLISHERS

Visit *The Learning Site!*
www.harcourtschool.com

Chapter 2 Weather and the Seasons

The Big Idea. vi
Quick and Easy Project 1
Insta-Labs 2
Vocabulary 4
Earth Science Content 6
CRCT Practice 20

Georgia Performance Standards in This Chapter

S1E1 Students will observe, measure, and communicate weather data to see patterns in weather and climate.

   a Identify different types of weather and the characteristics of each type.

   b Investigate weather by observing, measuring with simple weather instruments (thermometer, wind vane, rain gauge), and recording weather data (temperature, precipitation, sky conditions, and weather events) in a periodic journal or on a calendar seasonally.

   c Correlate weather data (temperature, precipitation, sky conditions, and weather events) to seasonal changes.

Chapter 3 Changes in Water

The Big Idea. 24
Quick and Easy Project 25
Insta-Labs 26
Vocabulary 28
Earth Science Content 30
CRCT Practice 40

Georgia Performance Standards in This Chapter

S1E2 Students will observe and record changes in water as it relates to weather.

   a Recognize changes in water when it freezes (ice) and when it melts (water).

   b Identify forms of precipitation such as rain, snow, sleet, and hailstones as either solid (ice) or liquid (water).

   c Determine that the weight of water before freezing, after freezing, and after melting stays the same.

   d Determine that water in an open container disappears into the air over time, but water in a closed container does not.

Chapter 4 Light and Sound

The Big Idea. 44

Quick and Easy Project 45

Insta-Labs 46

Vocabulary 48

Physical Science Content 50

CRCT Practice 64

Georgia Performance Standards in This Chapter

S1P1 Students will investigate light and sound.

a Recognize sources of light.

b Explain how shadows are made.

c Investigate how vibrations produce sound.

d Differentiate between various sounds in terms of (pitch) high or low and (volume) loud or soft.

e Identify emergency sounds and sounds that help us stay safe.

Chapter 5 Magnets

The Big Idea. 68

Quick and Easy Project 69

Insta-Labs 70

Vocabulary 72

Physical Science Content 74

CRCT Practice 84

Georgia Performance Standards in This Chapter

S1P2 Students will demonstrate effects of magnets on other magnets and other objects.

a Demonstrate how magnets attract and repel.

b Identify common objects that are attracted to a magnet.

c Identify objects and materials (air, water, wood, paper, your hand, etc.) that do not block magnetic force.

Chapter 6 All About Plants

The Big Idea. 88
Quick and Easy Project 89
Insta-Labs 90
Vocabulary 92
Life Science Content. 94
CRCT Practice 104

Georgia Performance Standards in This Chapter

S1L1 Students will investigate the characteristics and basic needs of plants and animals.

a Identify the basic needs of a plant.
1. Air
2. Water
3. Light
4. Nutrients

c Identify the parts of a plant—root, stem, leaf, and flower.

Chapter 7 All About Animals

The Big Idea. 108
Quick and Easy Project 109
Insta-Labs 110
Vocabulary 112
Life Science Content. 114
CRCT Practice 124

Georgia Performance Standards in This Chapter

S1L1 Students will investigate the characteristics and basic needs of plants and animals.

b Identify the basic needs of an animal.
1. Air
2. Water
3. Food
4. Shelter

d Compare and describe various animals—appearance, motion, growth, basic needs.

Weather and the Seasons

The Big Idea

You can observe, measure, and communicate to see patterns in weather.

On this page, show what you learn as you read this chapter.

Essential Question

What is weather?

Essential Question

How does weather change with each season?

Essential Question

How can we measure weather?

GO online ▶ Student eBook
www.hspscience.com

Cool Colors

What to Do

1. Put a thermometer inside each T-shirt.

2. Put the T-shirts in a sunny place. Record the temperature in each T-shirt.

3. Wait one hour. Record the temperatures again. Which colors stayed cooler? Which color got the warmest?

_____

_____

_____

_____

Draw Conclusions

What kinds of colors will help you stay cool?

_____

_____

_____

_____

Lesson 1

Observing Weather

1. Look out the window. Observe the sky. Observe what people are wearing.

2. What can you tell about the weather?

_____

_ _ _ _ _ _ _ _ _ _ _ _ _ _ _ _ _ _

_____

_ _ _ _ _ _ _ _ _ _ _ _ _ _ _ _ _ _

_____

3. Repeat each day for a week. Make a chart to show weather data.

New Leaves

1. Observe a plant stem or branch in spring.

2. Use a hand lens and a tape measure.

3. Observe the size, shape, and color of the new leaves.

4. How will the leaves change as they grow?

_____

– – – – – – – – – – – – – – – – – – –

_____

Observe Wind

1. Go outside.

2. Use soap bubbles to blow bubbles or hold up a pinwheel.

3. Which way is the wind blowing? How strong is the wind?

_____

– – – – – – – – – – – – – – – – – – –

_____

Vocabulary

 **Weather** is what the air outside is like. Rainy, sunny, windy, cloudy, and snowy are different kinds of weather.

 A **season** is a time of year. There are four seasons in a year—spring, summer, fall, and winter.

 **Temperature** is the measure of how hot or cold something is.

 A **thermometer** is a tool you use to measure temperature.

 A **wind vane** is a tool you use to measure the direction of the wind.

 A **rain gauge** is a tool you use to measure how much rain falls.

What is Weather?

sunny day

snowy day

Weather is how the air feels outside.
The air may be warm or cool.

1. _weather_ is how the air feels outside.

2. Circle the picture that shows sunny weather.

6

rainy day

windy day

Weather can change every day.

3. Put an **X** over the picture that shows windy weather.

4. Draw a picture showing rainy weather.

Types of Weather

cloudy and cool

There are different kinds of weather.
It may be cloudy and cool.

---

5. Look out the window.
Draw a picture of what the
weather looks like. Write
two words that describe
what the weather is like.

_____

_ _ _ _ _ _ _ _ _

sunny and cool

hot and sunny

Sunny weather warms the air.
A snowy day may be cloudy and cold.

6. Look at the picture. How can you tell that
the weather is hot and sunny?

_____

_____

_____

_____

_____

_____

How Does Weather Change with Each Season?

spring

summer

A **season** is a time of year.
Each year has four seasons.

1. A _____ is a time of year.

2. What are the four seasons?

_____

_____

winter

fall

Each season has different weather.
The weather pattern repeats each year.

3. Does spring always come after winter? Explain.

_____

_ _

_____

_____

_ _

_____

Spring

Spring comes after winter.

In spring, the weather gets warmer.
There may be many rainy days.

4. What kind of weather does spring have?

_____

— — — — — — — — — — — — — — — — — —

_____

Summer

Summer comes after spring.

In summer it may be hot and sunny.
Storms can change the weather.

---

5. Draw a picture showing an activity you can do during the summer.

Fall

Fall comes after summer.

Weather changes again in fall.
It may be sunny or cloudy and cool.

6. _____ is the season
that comes after summer.

7. What kind of weather does fall have?

Winter

Winter comes after fall.

Winter weather can be cold.
It snows in some places.

8. Why are the children wearing heavy clothing?

_____

_ _

_____

_ _

_____

How Can We Measure Weather?

We can measure rain.

We measure weather in many ways.
We can see patterns in the weather.

1. Why is it important to measure weather?

_____

_ _

_____

_ _

_____

Measuring Temperature

Temperatures can be hot in summer.

thermometer

Temperature is how hot or cold it is.
A **thermometer** shows the temperature.

_____
_ _ _ _ _ _ _ _ _ _ _ _

2. _____ is the measure of how hot or cold something is.

3. Circle the name of the tool that measures temperature.

balance   thermometer   hand lens

Measuring Wind

wind vane

A **wind vane** measures wind.
It shows which way the wind blows.

---

4. What does a wind vane measure?

_____

_ _ _ _ _ _ _ _ _ _ _ _ _ _ _ _

_____

5. What does a wind vane show?

_____

_ _ _ _ _ _ _ _ _ _ _ _ _ _ _ _

_____

Measuring Rain

rain gauge

A **rain gauge** measures rainfall.
It shows how much rain has fallen.

6. How does a rain gauge help you learn
   about weather?

   _____

   _ _ _ _ _ _ _ _ _ _ _ _ _ _ _ _ _

   _____

CRCT Practice

Fill in the circle in front of the letter of the best choice.

1. **What do the words rainy, snowy, and windy describe?**

   ○ A. kinds of rocks
   ○ B. kinds of weather
   ○ C. kinds of clouds

   S1E1a

2. **Which is the hottest season of the year?**

   ○ A. spring
   ○ B. winter
   ○ C. summer

   S1E1c

3. **Look at the picture. What season does it show?**

   ○ A. summer
   ○ B. fall
   ○ C. winter

   S1E1c

4. Which of these tools shows the direction of the wind?

○ A.

○ B.

○ C.

S1E1b

5. It is winter. Snow is falling. Which word describes the weather?

○ A. rainy
○ B. warm
○ C. cold

S1E1a

6. After months of cold weather, the temperature warms up. What season is it?

   O A. spring
   O B. summer
   O C. fall   **S1E1c**

7. Which of these tools measures temperature?

   O A. a rain gauge
   O B. a thermometer
   O C. a wind vane   **S1E1b**

8. **It is summer. The sun is shining. Which word best describes the weather?**

    ○ A. windy

    ○ B. cold

    ○ C. hot

    S1E1c

9. **Which of these words does NOT tell about weather?**

    ○ A. rainy

    ○ B. cloudy

    ○ C. happy

    S1E1a

10. **Which of these tools measures rainfall?**

    ○ A.

    ○ B.

    ○ C.

    S1E1b

3 Changes in Water

The Big Idea

You can observe and record changes in water, using the weather.

On this page, show what you learn as you read this chapter.

Essential Question

What are some forms of precipitation?

Essential Question

How can water change?

GO online

Student eBook
www.hspscience.com

Weight of Water

What to Do

1. Put some water in a zip-top plastic bag. Do not fill the bag completely.

2. Estimate the weight of the water in the bag before freezing, after freezing, and after melting.

3. Measure the water in the bag before freezing, after freezing, and after melting. Compare the measurements.

You need
- water
- zip-top plastic bag
- spring scale
- freezer

Draw Conclusions

1. Were your estimates correct?

_____

_ _ _ _ _ _ _ _ _ _ _ _ _ _ _ _ _ _ _

_____

2. What did you find out about water?

_____

_ _ _ _ _ _ _ _ _ _ _ _ _ _ _ _ _ _ _

_____

_____

_ _ _ _ _ _ _ _ _ _ _ _ _ _ _ _ _ _ _

_____

Lesson 1

Make a Rainbow

1. Place a mirror in a jar of water.

2. Turn off the classroom lights. Shine a flashlight on the mirror.

3. Move the flashlight around. Observe. What happens?

_____

_ _ _ _ _ _ _ _ _ _ _ _ _ _ _

_____

_____

_ _ _ _ _ _ _ _ _ _ _ _ _ _ _

_____

Evaporation

1. Get two jars. Put the same amount of water in each jar.

2. Mark the water line on each jar with a piece of tape. Cover one jar with a lid.

3. Put the jars in a warm, sunny place. Keep them there for several days.

4. Then compare the water in the jars. What happened? Why?

_____

_____

_____

_____

_____

_____

_____

_____

_____

_____

Vocabulary

 **Precipitation** is any form of water that falls from the sky. Rain, snow, sleet, and hail are all forms of precipitation.

 **Sleet** is frozen or partly frozen rain.

 **Hail** is balls of ice that fall from the sky.

 To **freeze** is to change from a liquid to a solid.

To **melt** is to change from a solid to a liquid.

To **evaporate** is to change from a liquid to a gas.

To **condense** is to change from a gas to a liquid.

The **water cycle** is the movement of water from Earth to the air and back again.

What Are Some Forms of Precipitation?

Snow is solid precipitation.

Precipitation is water that falls from the sky. Solid precipitation has its own shape.

1. What is precipitation?

_____

– – – – – – – – – – – – – – – – – – – –

_____

2. Look at the picture. What kind of precipitation do you see? What form is it in?

_____

– – – – – – – – – – – – – – – – – – – –

_____

Hail is solid precipitation.

Sleet is solid precipitation.

Tiny drops of ice are called **sleet**.
Big balls of ice are called **hail**.

3. How do you know that hail and sleet are both solids?

_____

_____

_____

_____

_____

Liquid Precipitation

Water can fall as rain.

Rain is liquid precipitation.
It does not have its own shape.

---

4. Circle the word that names what form of precipitation rain is.

gas   solid   liquid

5. How is rain different from snow, hail, and sleet?

_____

_ _ _ _ _ _ _ _ _ _ _ _ _ _ _ _ _ _ _ _

_____

The sun shines through rain.
This makes a rainbow.

6. Look at the picture. How was this rainbow made?

_____

_ _ _ _ _ _ _ _ _ _ _ _ _ _ _ _ _ _ _ _

_____

_____

_ _ _ _ _ _ _ _ _ _ _ _ _ _ _ _ _ _ _ _

_____

How Can Water Change?

Ice is a solid.

Water that gets cold enough **freezes**.
It changes to a solid.

1. What happens to water when it freezes?

_____

_ _

_____

2. What form of water is ice?

_____

_ _

_____

Water is a liquid.

Ice that gets warm enough **melts**.
It changes to a liquid.

---

3. What happens when ice melts?

_____

– – – – – – – – – – – – – – – – – – – –

_____

4. Look at the picture. Draw an **X** over the
liquid in the picture.

More Changes

Water in gas form is water vapor.

Water **evaporates** when it is heated.
It changes to a gas.

---

5. What happens to water that gets heated?

_____

_ _

_____

6. After water has been heated, it changes

_____

_ _ _ _ _ _ _ _ _ _ _

from a liquid to a _____.

Water vapor condenses on a cold glass

Water vapor **condenses** when cold.
It changes to drops of liquid.

7. Circle the word that tells what happens when water vapor gets cold.

8. Drops of liquid form on the outside of a cold glass on a hot day. What do you think would happen if the drops of liquid were heated up again?

_____

_ _

_____

The Water Cycle

2 Water vapor condenses and forms clouds.

1 Water evaporates.

Water on Earth evaporates.
It moves into the air.

---

9. What happens after water evaporates and moves into the air?

_____

_____

3 Rain or snow falls.

4 The cycle continues.

Then it moves back again.
That is the **water cycle**.

---

10. Does the water cycle end? Explain.

_____

_____

_____

_____

_____

CHAPTER 3

CRCT Practice

Fill in the circle in front of the letter of the best choice.

1. It is a winter day. The temperature is very cold. What will happen to the water in a puddle?

   ○ A. It will freeze.

   ○ B. It will melt.

   ○ C. It will condense.

   S1E2a

2. What kind of solid precipitation has fallen in this picture?

   ○ A. sleet

   ○ B. rain

   ○ C. snow

   S1E2b

3. In the water cycle, what happens after the sun warms the water?

   ○ A. The water flows.

   ○ B. The water evaporates.

   ○ C. It snows.

   S1E2d

4. Which of these is liquid precipitation?

    ○ A. hail
    ○ B. clouds
    ○ C. rain

S1E2b

5. Look at the picture. Rain has fallen on the house and yard. It is a warm summer day. What will happen to the water in the puddles?

    ○ A. It will freeze.
    ○ B. It will evaporate.
    ○ C. It will condense.

S1E2d

6. **What is happening to the ice cube in the picture?**

- ○ A. It is melting.
- ○ B. It is evaporating.
- ○ C. It is condensing.

S1E2a

7. **You have some water. The water freezes and turns into ice. How much will the ice weigh?**

- ○ A. It will weigh the same as the water.
- ○ B. It will weigh more than the water.
- ○ C. It will weigh less than the water.

S1E2c

8. It is a warm sunny day. What will happen to the water in an open container?

   ○ A. It will melt.

   ○ B. It will freeze.

   ○ C. It will evaporate. S1E2d

9. Which is another name for melted ice?

   ○ A. snow

   ○ B. water

   ○ C. water vapor S1E2a

10. Look at the picture. The ice in glass A weighs 10 ounces. The ice melts to make water in glass B. How much does the water in glass B weigh?

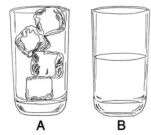

A   B

   ○ A. 12 ounces

   ○ B. 8 ounces

   ○ C. 10 ounces S1E2c

Light and Sound

The Big Idea

Light can make shadows.
Vibrations make sound.

On this page, show what you learn as you read this chapter.

Essential Question

What is light?

Essential Question

What is sound?

GO online ▶ Student eBook
www.hspscience.com

Investigate Pitch

You need
- 4 bottles
- water

What to Do

1. Blow across the open top of an empty bottle. Listen. Predict how the sound will change if you put water in the bottle.

2. Put a different amount of water in each bottle. Blow across the four bottles.

3. Which have a high pitch? Which have a low pitch? Arrange the bottles from highest pitch to lowest pitch.

_____

_____

_____

_____

Draw Conclusions

What effect does adding water have on the sound?

_____

_____

Lesson 1

What Can Light Pass Through?

1. Get some art materials.

2. Predict which ones light will pass through. Which ones will block light? Record your predictions.

Materials	Predictions

3. Test your ideas in a sunny place or next to a lamp. **CAUTION:** The lamp may get hot.

4. Were your predictions correct? What did you observe?

_____

_____

_____

_____

Straw Instruments

1. Cut a straw so the top forms a V.

2. Pinch the top with your lips. Blow very hard. Listen.

3. Then cut off some of the straw at the bottom. Blow again.

4. How does the sound change?

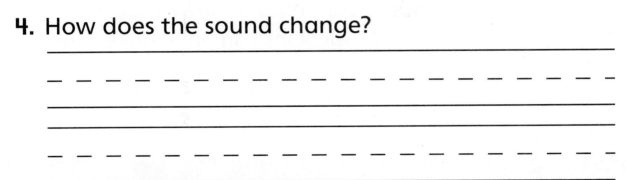

Vocabulary

 **Light** is a kind of energy that lets us see.

 A **shadow** is a dark place made when an object blocks light.

 **Sound** is a kind of energy that you hear. Musical instruments make sound.

A **vibration** is a movement back and forth. Strings on a guitar vibrate to make sound.

Volume is how loud or soft a sound is. A whisper is a soft sound.

Pitch is how high or low a sound is. Wind chimes can have a high pitch.

What Is Light?

light from the sun

Light is a kind of energy.
It helps us see.

1. What kind of energy lets us see?

   -

2. Where is the light in the picture coming from?

   -

light from a fire

light from a lamp

Most of our light comes from the sun.
We also get light from other things.

3. Name two things that give off light.

4. Draw a picture of each object.

Shadows

shadow

Objects block light.
That makes a **shadow**.

---

5. An object that blocks light

_____

– – – – – – – – – – – –

makes a _____.

We can see through glass.

Light can move through clear objects.
We can see through clear objects.

6. Put an **X** over the object that lets light move through it.

7. Can light move through the floor? Explain.

What Is Sound?

A whisper is a sound.

Sound is energy that we hear.
We hear sounds with our ears.

1. What kind of energy can we hear?

2. Look at the picture. Circle the body part that helps us hear sounds.

Guitars make sound.

The strings vibrate.

Something moving makes sound.
This motion is called **vibration**.

3. Put a box around the word that means motion makes sound.

4. Draw a picture of something that makes sound. Label your drawing.

Loud and Soft

soft

loud

There are loud sounds and soft sounds.
Volume is how loud or soft a sound is.

---

5. A fire truck makes a

_____

_ _ _ _ _ _ _ _ _ _

_____ sound.

_____

_ _ _ _ _ _ _ _ _ _ _

6. A whisper is a _____ sound.

When is the volume loudest?

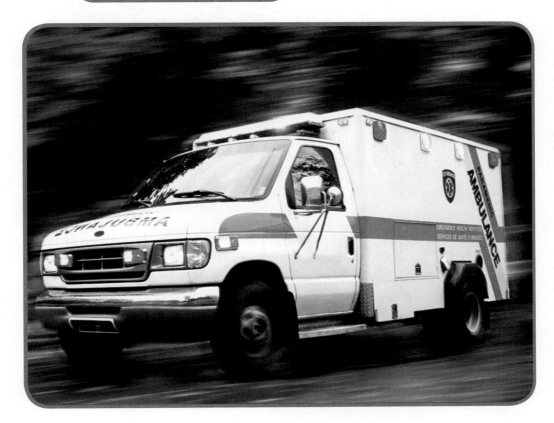

Close sounds are easier to hear.
Faraway sounds are harder to hear.

7. Look at the picture. Is it easier to hear the sounds that this object makes when it is close to you or far away from you? Explain.

_____

_ _ _ _ _ _ _ _ _ _ _ _ _ _ _ _ _ _ _ _

_____

_____

_ _ _ _ _ _ _ _ _ _ _ _ _ _ _ _ _ _ _ _

_____

High and Low

high pitch

low pitch

Some sounds are high.
Others are low.

8. Draw a picture of something that makes a high
sound or a low sound. Label your drawing.

Singers need to sing the right pitch.

tuning fork

Pitch tells how high or low a sound is. A tuning fork keeps the same pitch.

9. _____ tells how high or low a sound is.

Musical Instruments

drum

You can make music with instruments.
You hit a drum to make vibrations.

---

10. How does a drum make sound?

_____

- - - - - - - - - - - - - - - -

_____

_____

- - - - - - - - - - - - - - - -

_____

guitar

trumpet

You blow into a horn to make vibrations.
You pull on strings to make vibrations.

---

11. Draw an **X** over the part of the guitar that vibrates to make sound in the air.

Safety Sounds

Some sounds can keep us safe.
Alarms and horns warn us.

---

12. How can sounds keep us safe?

_____

- - - - - - - - - - - - - - - - -

_____

13. Name two sounds that can keep you safe.

_____

- - - - - - - - - - - - - - - - -

_____

Voices we hear also keep us safe.
Adults can tell us how to be safe.

14. Teachers can tell us how to be safe, too. List two rules that your teacher asks you to follow to keep you safe.

_____

_ _ _ _ _ _ _ _ _ _ _ _ _ _ _ _ _

_____

_ _ _ _ _ _ _ _ _ _ _ _ _ _ _ _ _

_____

CRCT Practice

Fill in the circle in front of the letter of the best choice.

1. **Which of these objects gives off light?**

   ○ A. a flashlight

   ○ B. a heater

   ○ C. a fan

   S1P1a

2. **What kind of pitch does the sound of a wind chime have?**

   ○ A. low

   ○ B. quiet

   ○ C. high

   S1P1d

3. **Why is the girl in the picture in a shadow?**

   ○ A. The sun shines through the window.

   ○ B. The building blocks the sun's light.

   ○ C. The building makes the street warm.

   S1P1b

4. Look at the picture. What causes the guitar string to make sound?

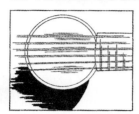

○ A. pitch

○ B. vibrations

○ C. loudness

5. Which sound is soft or quiet?

○ A. a drum

○ B. an airplane

○ C. a whisper

6. **Which of these objects will block light and make a shadow?**

   ○ A. a wooden door
   ○ B. a glass window
   ○ C. a piece of clear plastic   S1P1b

7. **Look at the picture. Which of these objects is a source of light?**

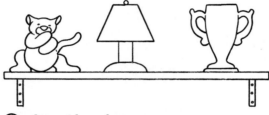

   ○ A. the bear
   ○ B. the trophy
   ○ C. the lamp   S1P1a

8. Look at the picture. Which makes the sound with the lowest pitch?

    ○ A. the bell

    ○ B. the drum

    ○ C. the whistle   **S1P1d**

9. What is made when an object blocks light?

    ○ A. a shadow

    ○ B. sunlight

    ○ C. sound   **S1P1b**

10. Which of these sounds helps keep us safe?

    ○ A. a piano

    ○ B. the wind blowing

    ○ C. a fire alarm   **S1P1e**

5 Magnets

The Big Idea Magnets can attract objects that have iron in them. They can pull iron objects through some materials.

On this page, show what you learn as you read this chapter.

Essential Question

What can a magnet do?

Essential Question

What can a magnet pull through?

GO online Student eBook
www.hspscience.com

Make a Magnetic Toy

What to Do

1. Cut a small kite from tissue paper.

2. Tie a paper clip to a thread. Tape the clip to the kite. Tape the end of the thread to a table.

3. Hold the magnet above the kite but not touching the kite. Use the magnet to make your kite fly.

Draw Conclusions

What part of the kite does the magnet pull? Why?

_____

– – – – – – – – – – – – – – – –

_____

_____

– – – – – – – – – – – – – – – –

_____

You need
- scissors
- tissue paper
- paper clip
- tape
- magnet
- thread

Lesson 1

Poles of a Magnet

1. Get two magnets.

2. Bring the N end of one magnet toward the S end of the other magnet.

3. Repeat, using the two N ends and then the two S ends.

4. What did you find out?

_____

_ _ _ _ _ _ _ _ _ _ _ _ _ _ _ _ _ _ _ _

_____

_____

_ _ _ _ _ _ _ _ _ _ _ _ _ _ _ _ _ _ _ _

_____

_____

_ _ _ _ _ _ _ _ _ _ _ _ _ _ _ _ _ _ _ _

_____

Move It with a Magnet

1. Find out through what materials a magnet can pull metal objects.

2. Get a strong magnet, a metal clip, paper, cloth, and other materials.

3. Try to attract the metal clip through the paper.

4. Now try to attract the metal clip through the cloth and then through the other materials.

5. Tell what you observe.

_____

_ _ _ _ _ _ _ _ _ _ _ _ _ _ _ _ _ _ _

_____

_____

_ _ _ _ _ _ _ _ _ _ _ _ _ _ _ _ _ _ _

_____

Vocabulary

A **magnet** is an object that will attract things made of iron.

A **pole** is the end of a magnet, where the pull is the strongest. A magnet has two poles—an N pole and an S pole.

Repel means to push away. Poles that are the same repel each other.

 **Attract** means to pull toward. Poles that are different attract each other.

 **Magnetic force** is the pulling force of a magnet. Some magnets are strong enough to pull iron objects through paper and other materials.

What Can a Magnet Do?

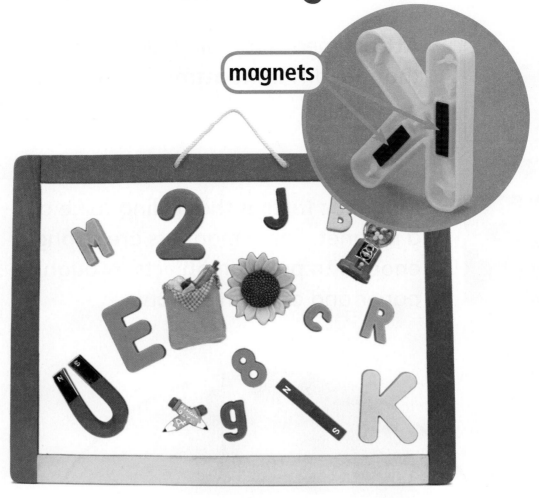

magnets

A **magnet** pulls things made of iron.

---

1. What object can pull things that have iron in them?

_____

- - - - - - - - - - - - - - - - - - -

_____

2. Circle the bar magnet.

3. Draw an **X** over the horseshoe magnet.

iron car

This car has iron in it.
A magnet can pull it.

4. Why can a magnet pull this car?

_____

_ _ _ _ _ _ _ _ _ _ _ _ _ _ _ _ _ _ _

_____

Objects Magnets Pull

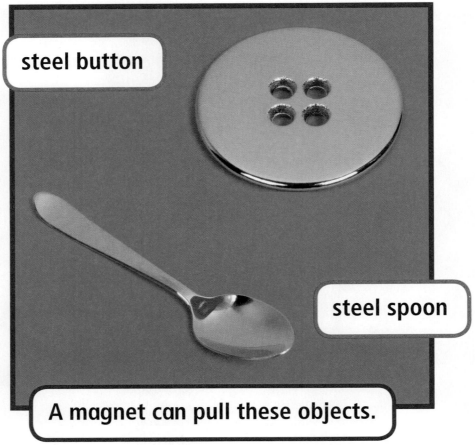

steel button

steel spoon

A magnet can pull these objects.

Steel has iron in it.
A magnet pulls it.

---

5. Name two objects that a magnet could pull.

_____

_ _

_____

6. Draw a picture of each object.

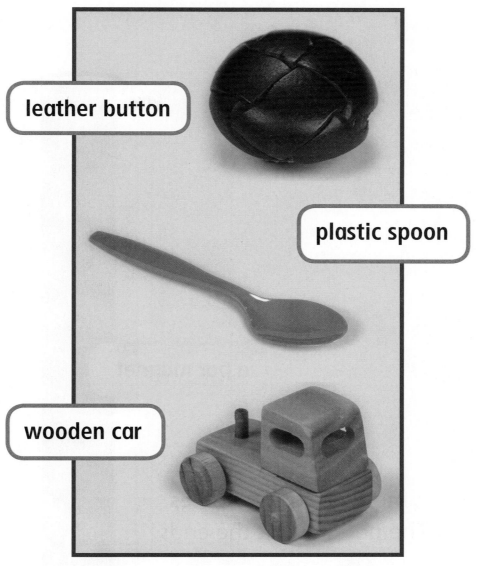

leather button

plastic spoon

wooden car

Magnets do not pull these things.
They are not made of iron or steel.

7. Circle what magnets could pull.

nail   cup   crayon

pencil   key   eraser

Poles of a Magnet

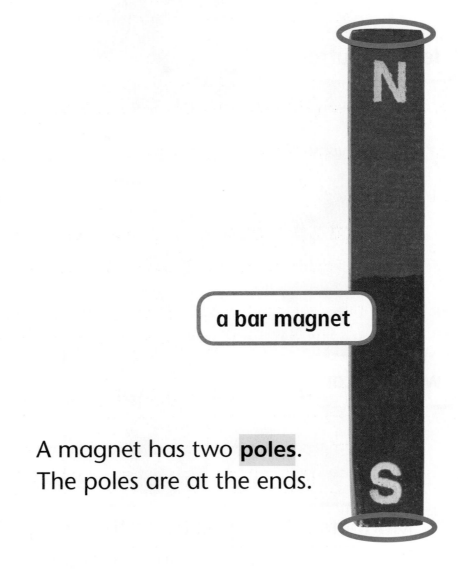

a bar magnet

A magnet has two **poles**.
The poles are at the ends.

8. The end of a magnet is called

   _____

   _ _ _ _ _ _ _ _ _ _

   the _____ .

9. A magnet has an S pole and an

   _____

   _ _ _ _ _ _ _ _ _ _

   _____ pole.

Magnets Repel

repel

Poles that are the same **repel**.
They push each other away.

---

10. Circle the word that means the same as
to push away.

Magnets Attract

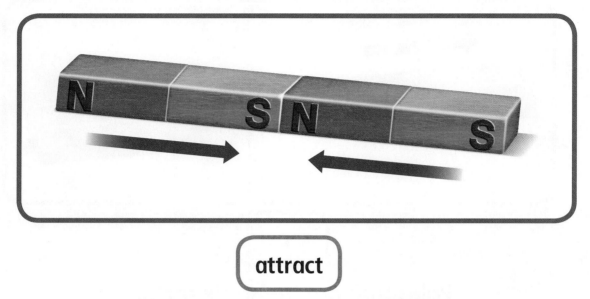

attract

Poles that are different **attract**.
They pull each other.

---

11. Underline the word that means the same as **to pull toward each other**.

12. What pole would attract the N pole of a magnet?

_____

_ _ _ _ _ _ _ _ _ _ _ _ _ _ _ _ _ _

_____

What Can a Magnet Pull Through?

A paper clip is on the kite. The magnet pulls it.

A magnet's pull is a force.

---

1. Draw a box around the place that shows the magnet's force.

What Magnets Pull Through

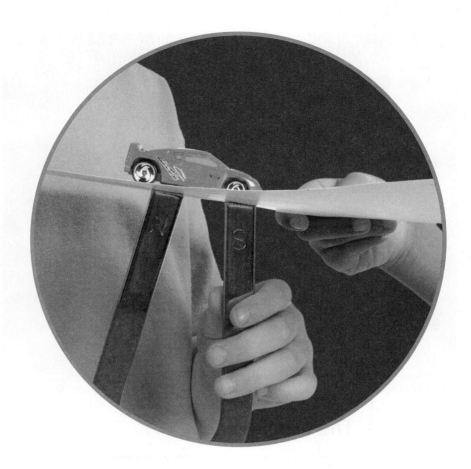

Some magnets are very strong.
They can pull through things.

2. What is the magnet in the picture pulling through?

_____

_ _ _ _ _ _ _ _ _ _ _ _ _ _ _ _ _ _

_____

> This magnet is strong.
> Its force pulls through water.

Magnets can even pull through wood.
A magnet's pull is called **magnetic force**.

_____

- - - - - - - - - - - - -

3. A magnet's pull is called its _____.

4. Why isn't the paper clip on the bottom of the cup?

_____

- - - - - - - - - - - - - - - - - - - -

_____

CHAPTER 5

CRCT Practice

Fill in the circle in front of the letter of the best choice.

1. What happens when you put together two of a magnet's poles that are not the same?

   ○ A. They will break.
   ○ B. They will repel each other.
   ○ C. They will attract each other. **S1P2a**

2. Angie is holding these objects. Which object will a magnet attract?

   ○ A. a metal paper clip
   ○ B. an eraser
   ○ C. a straw hat **S1P2b**

3. A magnet can not pull a toy through cardboard. What does this tell you?

   ○ A. The toy is a magnet.
   ○ B. They toy is not made of iron or steel.
   ○ C. The cardboard does not block the magnetic force. **S1P2a** **S1P2b** **S1P2c**

4. **Look at the picture. What will happen when these two magnets are put together?**

○ A. They will repel.

○ B. They will vibrate.

○ C. They will attract.

S1P2a

5. **Which of the following sentences about magnets is NOT true?**

○ A. Magnetic force can pull things through air.

○ B. Magnets attract things no matter what they are made of.

○ C. Magnetic force can pull things through water.

S1P2c

6. Look at the picture. Which object is attracted to the magnet?

- ○ A. the rock
- ○ B. the sailboat
- ○ C. the paper clip

S1P2b

7. Look at the picture. Which sentence is TRUE?

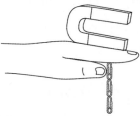

- ○ A. Wood blocks magnetic force.
- ○ B. Air blocks magnetic force.
- ○ C. Your hand does not block magnetic force.

S1P2c

8. Which two magnets will attract each other?

○ A.

○ B.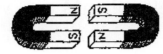

○ C. 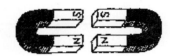   **S1P2a**

9. Which object will a magnet NOT attract?

○ A. a penny
○ B. a pencil
○ C. a key   **S1P2b**

10. A magnet will pull a paper clip through water. What does this tell you?

○ A. Water does not block magnetic force.
○ B. A paper clip is not attracted to a magnet.
○ C. The magnet is not strong.   **S1P2c**

6 All About Plants

The Big Idea

Plants need air, water, light, and nutrients to live and grow. Different parts of plants help plants get what they need.

On this page, show what you learn as you read this chapter.

Essential Question

What do plants need?

Essential Question

What are the parts of plants?

GO online Student eBook
www.hspscience.com

Plants' Needs

You need
- 3 different kinds of plants

What to Do

1. Do all plants need exactly the same things?

   _____

   - - - - - - - - - - -

   _____

2. Get three plants. Plan an investigation to answer the question.

3. Follow your plan.

4. Draw pictures, and write sentences to record what happens.

Draw Conclusions

1. Did you find an answer to the question?

   _____

   - - - - - - - - - - - - - - - - - -

   _____

2. If you did not find an answer, how could you change your investigation?

   _____

   - - - - - - - - - - - - - - - - - -

   _____

89

Make a Model Plant

1. Get paper, clay, craft sticks, and other art materials.

2. Use them to make a model of a plant.

3. Then tell about what a real plant needs to live.

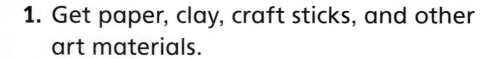

Lesson 2

How Roots Help

1. Push a craft stick deep into clay.

2. Push another craft stick into clay just a little.

3. Tap the side of each stick. What happens?

_____

_ _

_____

_____

_ _

_____

4. How is the first stick like a plant with roots?
How do roots hold a plant in place?

_____

_ _

_____

_____

_ _

_____

_____

_ _

_____

Vocabulary

 **Sunlight** is light that comes from the sun.

 **Nutrients** are minerals in the soil that plants need to grow and stay healthy.

 The **roots** are the parts of the plant that hold it in the soil and take in water and nutrients.

 The **stem** is the part of the plant that holds up the plant and lets food and water move through the plant.

The **leaves** are the parts of a plant that take in light and air and make food.

The **flowers** are the parts of a plant that make fruits.

The **fruits** are the parts of a plant that hold the seeds.

The **seeds** are the parts of a plant from which new plants grow.

What Do Plants Need?

sunlight

Plants make their own food.
They need **sunlight** to make food.

1. How do plants get food?

_____
_ _ _ _ _ _ _ _ _ _ _ _ _ _ _ _
_____

2. What is one thing that plants need to make food?

_____
_ _ _ _ _ _ _ _ _ _ _ _ _ _ _ _
_____

air

Plants also need air to make food.

3. Name another thing that plants need to
   make food.

   _____

   _ _ _ _ _ _ _ _ _ _ _ _ _ _ _ _ _ _ _

   _____

Water

water

Plants need water to make food.

4. Why do plants need water?

_____

- - - - - - - - - - - - - - - - -

_____

5. Where do plants get water?

_____

- - - - - - - - - - - - - - - - -

_____

Nutrients

soil

Plants need **nutrients** from soil, too.

6. What would happen if a plant did not get sunlight, air, water, and nutrients?

_____

_____

_____

_____

_____

What Are the Parts of Plants?

Plants have different parts.

flower

leaf

stem

roots

---

1. Name two parts that most plants have.

_____

\- \- \- \- \- \- \- \- \- \- \- \- \- \- \- \- \-

_____

Roots

roots

Roots take in water and nutrients.
Roots also help hold a plant in soil.

2. Look at the picture. Describe the kind of
roots the plant has.

_____

_____

_____

_____

Stems

stem

Stems hold up the plant.
Food and water move through stems.

3. Name two ways that a stem helps plants.

_____

_ _ _ _ _ _ _ _ _ _ _ _ _ _ _ _ _ _ _ _

_____

_____

_ _ _ _ _ _ _ _ _ _ _ _ _ _ _ _ _ _ _ _

_____

Leaves

leaves

Leaves take in light and air.
Leaves use light and air to make food.

4. Do all leaves look the same? Explain.

_____

_ _

_____

_ _

_____

Flowers, Fruits, and Seeds

Many plants have **flowers**.
Flowers make **fruits**.

flowers

5. Name the two parts of a plant pictured on this page.

_____

6. What do flowers do?

_____

fruits

seed

Seeds grow in the fruit.
New plants may grow from the seeds.

7. Circle the name of the part that grows in the fruit.

roots   seeds   flowers

8. What can grow from seeds?

_____

– –

_____

Fill in the circle in front of the letter of the best choice.

1. **Look at the picture. Which part of the plant holds it up?**

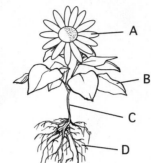

   ○ A. part B
   ○ B. part C
   ○ C. part D

   S1L1c

2. **Why does a plant need light?**

   ○ A. to make water
   ○ B. to get air
   ○ C. to make food

   S1L1a

3. **Which part of a plant makes fruits?**

   ○ A. flower
   ○ B. leaves
   ○ C. roots

   S1L1c

4. **Jim wants to grow a tomato plant. What does his plant need to grow and stay healthy?**

   ○ A. light, air, water, and nutrients

   ○ B. food and water

   ○ C. flowers, leaves, and fruits   **S1L1a**

5. **Look at the picture. What part of the plant does the arrow point to?**

   ○ A. stem

   ○ B. seed

   ○ C. flowers   **S1L1c**

6. **Look at the picture. How does this plant part help the plant?**

○ A. It holds up the plant.

○ B. It takes in nutrients.

○ C. It takes in sunlight.   **S1L1c**

7. **Where do plants get the water they need to grow?**

○ A. the sun

○ B. the air

○ C. the soil   **S1L1a**

8. **Look where the arrow is pointing. What does this part help the plant do?**

- ○ A. It holds the plant in the soil.
- ○ B. It makes food for the plant.
- ○ C. It takes in light.

S1L1c

9. **How do sunlight, air, nutrients, and water help plants?**

- ○ A. They help plants move.
- ○ B. They help plants live and grow.
- ○ C. They help plants learn.

S1L1a

10. **What do a plant's flowers do?**

- ○ A. They make roots.
- ○ B. They make fruits.
- ○ C. They hold it up.

S1L1c

All About Animals

The Big Idea

Animals need air, water, food, and shelter to live and grow. You can compare animals.

On this page, show what you learn as you read this chapter.

Essential Question

What do animals need?

Essential Question

How can we compare animals?

Go online ▶ Student eBook
www.hspscience.com

Which Foods Birds Eat

You need
- 2 foil pie plates
- bread crumbs
- chopped apples and grapes

What to Do

1. Put bread crumbs in one pie plate. Put fruit in the other.

2. Put both plates on a table outside.

3. Observe the birds that eat from each plate. Draw pictures to record what you observe.

Draw Conclusions

Do different birds eat different foods? How do you know?

_____

- - - - - - - - - - - - - - -

_____

_____

- - - - - - - - - - - - - - -

_____

Insta-Lab

Pet Food Survey

1. List some pet foods.

Kinds of Food	How Many

2. Then ask your classmates what their pets eat.

3. Make a tally mark next to each food.

4. Which food do the most pets eat?

_____

_____

_____

Make a Model

1. Make a model of an animal. Use chenille sticks.

2. Ask a classmate to guess the animal you made.

3. Tell why you made the model as you did.

_____

_ _ _ _ _ _ _ _ _ _ _ _ _ _ _ _

_____

_____

_ _ _ _ _ _ _ _ _ _ _ _ _ _ _ _

_____

Vocabulary

A **shelter** is a place where a person or an animal can be safe. Some owls use a hole in a tree for shelter.

An **appearance** is what something looks like. A porcupine has quills.

Growth is an increase in size.

What Do Animals Need?

Some animals eat plants.

panda

Animals need food to live and grow.

1. What do all animals need to live and grow?

_____

_ _ _ _ _ _ _ _ _ _ _ _ _ _ _ _ _ _

_____

2. What kind of food do pandas eat?

_____

_ _ _ _ _ _ _ _ _ _ _ _ _ _ _ _ _ _

_____

horse

deer

Animals drink water.

Animals need water, too.

3. Name another thing animals need to live and grow.

_____

– – – – – – – – – – – – – – – – –

_____

4. Where do animals get water?

_____

– – – – – – – – – – – – – – – – –

_____

Animals Need Air

Fish have gills.

Porcupines have lungs.

Animals need air.
Animals get air with lungs or gills.

5. Why do animals need air?

_____

- - - - - - - - - - - - - - - - -

_____

6. Name two body parts that help different animals get air.

_____

- - - - - - - - - - - - - - - - -

_____

Animals Need Shelter

owl

Foxes dig holes for shelter.

A **shelter** is a place to stay safe.
Many animals need shelter to live.

---

7. Draw a picture of an animal in a shelter.

How Can We Compare Animals?

Fish have scales.

Polar bears have fur.

Each animal has an **appearance**.
Appearance is the way something looks.

---

1. Circle the word that means "the way something looks."

2. Compare the fish and the polar bears in the pictures. How are they different?

_____

_ _ _ _ _ _ _ _ _ _ _ _ _ _ _ _ _

_____

_ _ _ _ _ _ _ _ _ _ _ _ _ _ _ _ _

_____

Compare How Animals Move

run

swim

fly

jump

Each animal moves in its own way.
Look at ways these animals move.

---

3. Think about your favorite animal. Draw a picture
showing how it moves.

Compare How Animals Grow

> **Puppies look like small dogs.**

All animals grow and change.
This is called **growth**.

---

4. Circle the word that tells about animals getting larger.

5. Look at the picture. How are the puppies and the adult dog alike?

_____

_ _

_____

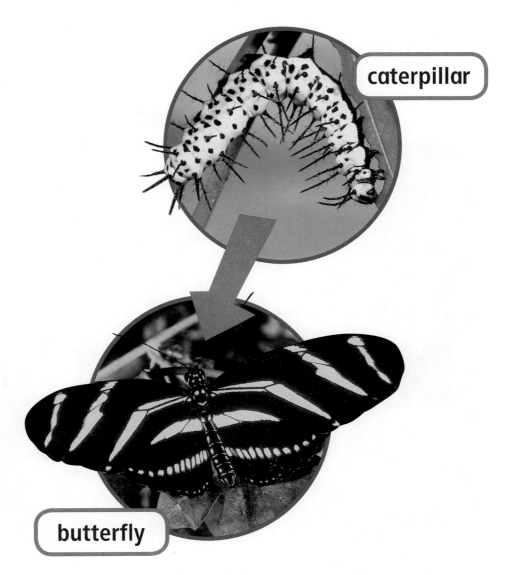

caterpillar

butterfly

Some animals look like their parents.
Others do not look like their parents.

6. Look at the pictures. How are the butterfly
and the caterpillar different?

_____

_ _

_____

_____

_ _

_____

Compare the Needs of Animals

raccoon

frog

All animals need food, water, and air.
Most animals need shelter.

---

7. Draw a picture of an animal meeting its needs.

elephants

Some animals may need a lot of food and water. Others may need only a little.

8. Do you think an elephant would need more food and water than a raccoon? Why or why not?

_____

_ _ _ _ _ _ _ _ _ _ _ _ _ _ _ _ _

_____

_____

_ _ _ _ _ _ _ _ _ _ _ _ _ _ _ _ _

_____

CRCT Practice

Fill in the circle in front of the letter of the best choice.

1. **Look at the picture. How is this animal using the hole in the tree?**

   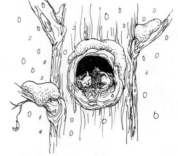

   ○ A. as water
   ○ B. as food
   ○ C. as shelter

   S1L1b

2. **Why do animals need food and water?**

   ○ A. to live and grow
   ○ B. to get air
   ○ C. as shelter

   S1L1b

3. **How do fish use gills?**

   ○ A. to swim
   ○ B. to get air
   ○ C. to find food

   S1L1b

4. **Look at the picture. How does this animal move?**

  ○ A. It flies.

  ○ B. It swims.

  ○ C. It crawls.   **S1L1d**

5. **How are penguins, owls, and eagles alike?**

  ○ A. They have fur.

  ○ B. They breathe with gills.

  ○ C. They have feathers.   **S1L1d**

6. **What do all animals need to live and grow?**

   ○ A. food, light, nutrients, and water

   ○ B. air, water, food, and shelter

   ○ C. light, air, water, and nutrients   **S1L1b**

7. **Look at the picture. How can you describe this animal?**

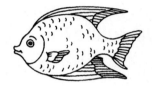

   ○ A. It can move from water to land.

   ○ B. It has scales and gills.

   ○ C. It has wings and feathers.   **S1L1d**

8. **Look at the picture. What will this animal look like when it is an adult?**

   O A. a frog
   O B. a fish
   O C. a butterfly

   S1L1d

9. **How would you describe a bear?**

   O A. It has scaly skin.
   O B. It has fur.
   O C. It has smooth, wet skin.

   S1L1d

10. **Jennifer saw a frog and a rabbit. How can she compare them?**

   O A. They move the same.
   O B. They both live in water.
   O C. They both have fur.

   S1L1d